The Ornamental Files

A Record of Flowers, Foliage, Colors, Fragrances, and Cherished Garden Memories

Property of:

ISBN: 978-1-968971-27-4

The desert and the parched land will be glad; the wilderness will rejoice and blossom. Like the crocus, 2 it will burst into bloom; it will rejoice greatly and shout for joy. Isaiah 35: 1-2

Ironstone
Tennessee

Disclaimer and Terms of Use:

The author and publisher have made every effort to ensure the completeness of this record book. However, they do not guarantee that all information is error-free or entirely comprehensive.

While diligent efforts have been made to verify the content, the author and publisher are not responsible for any errors, omissions, or differing interpretations of the material presented. Any unintended references that may appear to slight individuals, businesses, or organizations are purely coincidental.

Copies of this book are available for book signings, special events, and retail gift shop purchases. For direct ship information, of copies of 10 or more, please contact the author at ironstone_tennessee@yahoo.com.

The Ornamental Files
A Record of Flowers, Foliage, Colors, Fragrances, and Cherished Garden Memories

ISBN:978-1-968971-27-4

Printed in the United States of America.

First printing, 2025

Ironstone Tennessee

Calhoun Tennessee, 37309

www.ironstone-tennessee.com

Notes

Table of Contents

How to Use The Ornamental Plant Files

I want to thank you so much for choosing to purchase The Ornamental Plant Files. I welcome feedback and suggestions for ways to make it more useful for future editions.

There are more than a dozen page styles, each grouped together so that you can organize your plant files in the way that works best for you. When you're finished, you'll have created a valuable reference as you go forward, and a priceless heirloom to pass on to future generations of gardeners.

Important Dates and Data:

Record your **USDA growing zone, average and observed frost dates, average annual rainfall, soil type and pH here.** This information allows you to choose plants that will be most likely to do well in your climate. Most seed packets, plant labels and catalog listings will tell you the plant's range of growing zones.

Knowing your **average first and last frost dates** will help you to schedule appropriate shipping dates when purchasing live plants and trees online. It will also help you to schedule when to start seeds indoors, when it is (probably) safe to transplant outdoors, and when you will need to have raised beds and garden plots prepared for Spring and Fall planting.

Your **annual average precipitation** (snow and rain) will help you to choose what plants to grow based on your desired amount of time spent in the garden (and how big of a water bill you can afford). If you are a lazy gardener like me, or just don't have time in for your schedule for frequent watering, then choosing plants that can thrive with only rainfall (most of the time) makes sense.

Your **soil type and pH** will determine what amendments you may need, what gardening methods will be most successful for you, and what plants will and won't do well for you in the garden. This knowledge will same you a lot of time and money.

My Gardening Challenges:

These are factors that you can't change easily (or at all), such as a short growing season, your work schedule, soil type, physical limitations, small budget or climate. While it's easy to think of these challenges as a big negative, I would encourage you to instead think of them as boundaries that help you to keep your focus. Knowing from the start that you can't do All. The. Things, you are free to focus instead on the things you can do, and then make a plan to do those things well. Which brings us to:

My Gardening Goals:

Are you looking to build a flower garden from scratch? Do you want to replicate the look of a cottage garden? Do you want to surround yourself with the kind of flowers that your great-grandmother grew? Maybe you want to focus your attention on container flowers. Maybe your goal is to see how many different flowers and shrubs you can acquire for free, or how many new gardeners you can share your perennials with. Having goals and keeping them in focus will help you to achieve the garden of your dreams without going broke in the process.

Planting and Bloom Times:

Charts are provided to help you keep track of the times of year when you will be busy planting your ornamentals. For Bloom Times, the months are divided by weeks so that you can see when your flowers will bloom and when your trees and shrubs will put on Fall and Winter color.

Sections:

While the illustrations for each section may suggest a certain type of flower, tree or shrub, you are free to organize your plant files in whatever way makes sense to you. You might decide that each section represents a flower bed rather than a type of plant. Either way, there is space under each section number to write in a descriptive name such as (annuals, hostas, flowering trees etc.)

Common Name:

Just as it suggests, this is the name by which most people recognize a plant. Examples would be rose, iris, pansy, or boxwood.

Latin name:

This is the name that horticulturists use to refer to the plant. Often, there are multiple plants that have the same or a similar common name such as "bleeding heart". To most gardeners, this refer to the shade-loving dicentra plant which features red, pink, white or occasionally yellow heart-shaped blossoms that hang along a common stem. However to some, it refers to the fuchsias that are often sold as annuals in hanging baskets. Both plants produce absolutely beautiful flowers, but they are not the same things. If you know the Latin name, you can be sure you are getting the correct plant when ordering online or looking at plants at the garden center.

Cultivar:

Many plants come in a variety of sizes, shapes, fragrances, growth habits and colors. These plants usually are given a cultivar name to differentiate one from another. For example: There are many different climbing roses available. If I wanted a deep pink, shade-tolerant fragrant climber, I might choose "Zephirine Drouhin". "Julia Child", while it is fragrant, is yellow and not a climber, and would not suit my purpose.

Type:

This might refer to growth habit such as Climber, rambler, miniature, or shrub for roses, or tall bearded, dwarf, Dutch or Siberian for iris.

Bloom Time:

This can refer to the part of the season when the plant blooms such as early, mid or late spring for bulbs like daffodils, hyacinths or tulips. Or it can refer to plants that bloom more than once in the course of the growing season such as some iris and roses.

Fragrance:

Some flowers have no fragrance while others smell very strongly. Knowing the fragrance characteristics of plants can help you decide whether to grow them.

Source:

It's good to record where your plants come from for several reasons. If you decide you want more, you know where to go. If you aren't happy with a plant's performance, you'll know to try a different source next time. If someone asks you where you got that beautiful yellow peony, you can, if you choose, share your source.

Light Preference:

Some plants are very picky about how much sun or shade they need to thrive, while others are very flexible. If you're thinking about transplanting something, knowing its light requirement can help you to decide where to move it.

Soil Type:

Some plants are very picky about the soil type they prefer. If your favorite plant doesn't favor your soil, you will need to decide just how far you are willing to go to have that plant. Your soil can be improved, but it takes time and work.

Water Requirements:

Hydrangeas (as their name suggests) are usually gluttons for water. Knowing this can help you decide where to plant them (under the eaves of your house, or near the garden hose). It can also help you decide what to plant. If you live in a drought-prone area and you pay for water, you might not want to plant those hydrangeas at all.

Bloom Characteristics:

Here is the space to record the plant's blossom size, color, petal count, repeat blooming, frost tolerance, whether it is edible or toxic etc.

Soil Amendments:

Some plants require the application of fertilizers or pH adjustment from time to time. Record the plant's needs here so you can find the information quickly.

Diseases & Control:

Certain cultivars of plants are more disease prone than others. When you know your plants, you're better able to keep watch and head off a small problem before it becomes a large problem. You're also able to decide whether the plants that are most susceptible to disease are worth your time to grow.

Pests & Control:

Most plants have some kind of insect or animal that loves to feast on them. If you know what to watch out for, you can make a plan to prevent, treat or protect your plants as you see fit. You can also decide whether not it's worthwhile to grow those plants that are more vulnerable to pests.

Notes:

The notes area is for your observations, ideas, plans, recollections and thoughts you don't want to forget. Things like "this is my very favorite pansy, grow more next year" or "Grandma Boyd grew these" or "These took over the flower bed, DO NOT plant again".

I wish you Happy Gardening

Important Dates and Data

Growing Zone	Soil Type	Soil pH	Avg. 1st Frost	Avg. Last Frost	Avg. Annual Rainfall

My Gardening Challenges:

My Gardening Goals:

Planting Times

Month/Week	January				February				March			
Plant	**1**	**2**	**3**	**4**	**1**	**2**	**3**	**4**	**1**	**2**	**3**	**4**

Planting Times

Month/Week	April				May				June			
Plant	**1**	**2**	**3**	**4**	**1**	**2**	**3**	**4**	**1**	**2**	**3**	**4**

Planting Times

Month/Week	July				August				September			
Plant	1	2	3	4	1	2	3	4	1	2	3	4

Planting Times

Month/Week	October				November				December			
Plant	**1**	**2**	**3**	**4**	**1**	**2**	**3**	**4**	**1**	**2**	**3**	**4**

January Blooms

Week 1

Week 2

Week 3

Week 4

February Blooms

Week 1

Week 2

Week 3

Week 4

March Blooms

Week 1

Week 2

Week 3

Week 4

April Blooms

Week 1

Week 2

Week 3

Week 4

May Blooms

Week 1

Week 2

Week 3

Week 4

June Blooms

Week 1

Week 2

Week 3

Week 4

July Blooms

Week 1

Week 2

Week 3

Week 4

August Blooms

Week 1

Week 2

Week 3

Week 4

September Blooms

Week 1

Week 2

Week 3

Week 4

October Blooms

Week 1

Week 2

Week 3

Week 4

November Blooms

Week 1

Week 2

Week 3

Week 4

December Blooms

Week 1	Week 2

Week 3	Week 4

Notes

Section 1:

Ornamental Plant Files

Common Name	
Latin Name	
Cultivar	
Type	
Bloom Time	
Fragrance	
Source	
Light Preference	
Soil Type	
Water Requirement	
Bloom Characteristics	
Soil Amendments	
Diseases & Control	
Pests & Control	

Notes:

Ornamental Plant Files

Common Name	
Latin Name	
Cultivar	
Type	
Bloom Time	
Fragrance	
Source	
Light Preference	
Soil Type	
Water Requirement	
Bloom Characteristics	
Soil Amendments	
Diseases & Control	
Pests & Control	

Notes:

Ornamental Plant Files

Common Name	
Latin Name	
Cultivar	
Type	
Bloom Time	
Fragrance	
Source	
Light Preference	
Soil Type	
Water Requirement	
Bloom Characteristics	
Soil Amendments	
Diseases & Control	
Pests & Control	

Notes:

Ornamental Plant Files

Common Name	
Latin Name	
Cultivar	
Type	
Bloom Time	
Fragrance	
Source	
Light Preference	
Soil Type	
Water Requirement	
Bloom Characteristics	
Soil Amendments	
Diseases & Control	
Pests & Control	

Notes:

Ornamental Plant Files

Common Name	
Latin Name	
Cultivar	
Type	
Bloom Time	
Fragrance	
Source	
Light Preference	
Soil Type	
Water Requirement	
Bloom Characteristics	
Soil Amendments	
Diseases & Control	
Pests & Control	

Notes:

Ornamental Plant Files

Common Name	
Latin Name	
Cultivar	
Type	
Bloom Time	
Fragrance	
Source	
Light Preference	
Soil Type	
Water Requirement	
Bloom Characteristics	
Soil Amendments	
Diseases & Control	
Pests & Control	

Notes:

Ornamental Plant Files

Common Name	
Latin Name	
Cultivar	
Type	
Bloom Time	
Fragrance	
Source	
Light Preference	
Soil Type	
Water Requirement	
Bloom Characteristics	
Soil Amendments	
Diseases & Control	
Pests & Control	

Notes:

Ornamental Plant Files

Common Name	
Latin Name	
Cultivar	
Type	
Bloom Time	
Fragrance	
Source	
Light Preference	
Soil Type	
Water Requirement	
Bloom Characteristics	
Soil Amendments	
Diseases & Control	
Pests & Control	

Notes:

Ornamental Plant Files

Common Name	
Latin Name	
Cultivar	
Type	
Bloom Time	
Fragrance	
Source	
Light Preference	
Soil Type	
Water Requirement	
Bloom Characteristics	
Soil Amendments	
Diseases & Control	
Pests & Control	

Notes:

Ornamental Plant Files

Common Name	
Latin Name	
Cultivar	
Type	
Bloom Time	
Fragrance	
Source	
Light Preference	
Soil Type	
Water Requirement	
Bloom Characteristics	
Soil Amendments	
Diseases & Control	
Pests & Control	

Notes:

Ornamental Plant Files

Common Name	
Latin Name	
Cultivar	
Type	
Bloom Time	
Fragrance	
Source	
Light Preference	
Soil Type	
Water Requirement	
Bloom Characteristics	
Soil Amendments	
Diseases & Control	
Pests & Control	

Notes:

Ornamental Plant Files

Common Name	
Latin Name	
Cultivar	
Type	
Bloom Time	
Fragrance	
Source	
Light Preference	
Soil Type	
Water Requirement	
Bloom Characteristics	
Soil Amendments	
Diseases & Control	
Pests & Control	

Notes:

Ornamental Plant Files

Common Name	
Latin Name	
Cultivar	
Type	
Bloom Time	
Fragrance	
Source	
Light Preference	
Soil Type	
Water Requirement	
Bloom Characteristics	
Soil Amendments	
Diseases & Control	
Pests & Control	

Notes:

Ornamental Plant Files

Common Name	
Latin Name	
Cultivar	
Type	
Bloom Time	
Fragrance	
Source	
Light Preference	
Soil Type	
Water Requirement	
Bloom Characteristics	
Soil Amendments	
Diseases & Control	
Pests & Control	

Notes:

Notes

Section 2:

Ornamental Plant Files

Common Name	
Latin Name	
Cultivar	
Type	
Bloom Time	
Fragrance	
Source	
Light Preference	
Soil Type	
Water Requirement	
Bloom Characteristics	
Soil Amendments	
Diseases & Control	
Pests & Control	

Notes:

Ornamental Plant Files

Common Name	
Latin Name	
Cultivar	
Type	
Bloom Time	
Fragrance	
Source	
Light Preference	
Soil Type	
Water Requirement	
Bloom Characteristics	
Soil Amendments	
Diseases & Control	
Pests & Control	

Notes:

Ornamental Plant Files

Common Name	
Latin Name	
Cultivar	
Type	
Bloom Time	
Fragrance	
Source	
Light Preference	
Soil Type	
Water Requirement	
Bloom Characteristics	
Soil Amendments	
Diseases & Control	
Pests & Control	

Notes:

Ornamental Plant Files

Common Name	
Latin Name	
Cultivar	
Type	
Bloom Time	
Fragrance	
Source	
Light Preference	
Soil Type	
Water Requirement	
Bloom Characteristics	
Soil Amendments	
Diseases & Control	
Pests & Control	

Notes:

Ornamental Plant Files

Common Name	
Latin Name	
Cultivar	
Type	
Bloom Time	
Fragrance	
Source	
Light Preference	
Soil Type	
Water Requirement	
Bloom Characteristics	
Soil Amendments	
Diseases & Control	
Pests & Control	

Notes:

Ornamental Plant Files

Common Name	
Latin Name	
Cultivar	
Type	
Bloom Time	
Fragrance	
Source	
Light Preference	
Soil Type	
Water Requirement	
Bloom Characteristics	
Soil Amendments	
Diseases & Control	
Pests & Control	

Notes:

Ornamental Plant Files

Common Name	
Latin Name	
Cultivar	
Type	
Bloom Time	
Fragrance	
Source	
Light Preference	
Soil Type	
Water Requirement	
Bloom Characteristics	
Soil Amendments	
Diseases & Control	
Pests & Control	

Notes:

Ornamental Plant Files

Common Name	
Latin Name	
Cultivar	
Type	
Bloom Time	
Fragrance	
Source	
Light Preference	
Soil Type	
Water Requirement	
Bloom Characteristics	
Soil Amendments	
Diseases & Control	
Pests & Control	

Notes:

Ornamental Plant Files

Common Name	
Latin Name	
Cultivar	
Type	
Bloom Time	
Fragrance	
Source	
Light Preference	
Soil Type	
Water Requirement	
Bloom Characteristics	
Soil Amendments	
Diseases & Control	
Pests & Control	

Notes:

Ornamental Plant Files

Common Name	
Latin Name	
Cultivar	
Type	
Bloom Time	
Fragrance	
Source	
Light Preference	
Soil Type	
Water Requirement	
Bloom Characteristics	
Soil Amendments	
Diseases & Control	
Pests & Control	

Notes:

Ornamental Plant Files

Common Name	
Latin Name	
Cultivar	
Type	
Bloom Time	
Fragrance	
Source	
Light Preference	
Soil Type	
Water Requirement	
Bloom Characteristics	
Soil Amendments	
Diseases & Control	
Pests & Control	

Notes:

Ornamental Plant Files

Common Name	
Latin Name	
Cultivar	
Type	
Bloom Time	
Fragrance	
Source	
Light Preference	
Soil Type	
Water Requirement	
Bloom Characteristics	
Soil Amendments	
Diseases & Control	
Pests & Control	

Notes:

Ornamental Plant Files

Common Name	
Latin Name	
Cultivar	
Type	
Bloom Time	
Fragrance	
Source	
Light Preference	
Soil Type	
Water Requirement	
Bloom Characteristics	
Soil Amendments	
Diseases & Control	
Pests & Control	

Notes:

Ornamental Plant Files

Common Name	
Latin Name	
Cultivar	
Type	
Bloom Time	
Fragrance	
Source	
Light Preference	
Soil Type	
Water Requirement	
Bloom Characteristics	
Soil Amendments	
Diseases & Control	
Pests & Control	

Notes:

Notes

Section 3:

Ornamental Plant Files

Common Name	
Latin Name	
Cultivar	
Type	
Bloom Time	
Fragrance	
Source	
Light Preference	
Soil Type	
Water Requirement	
Bloom Characteristics	
Soil Amendments	
Diseases & Control	
Pests & Control	

Notes:

Ornamental Plant Files

Common Name	
Latin Name	
Cultivar	
Type	
Bloom Time	
Fragrance	
Source	
Light Preference	
Soil Type	
Water Requirement	
Bloom Characteristics	
Soil Amendments	
Diseases & Control	
Pests & Control	

Notes:

Ornamental Plant Files

Common Name	
Latin Name	
Cultivar	
Type	
Bloom Time	
Fragrance	
Source	
Light Preference	
Soil Type	
Water Requirement	
Bloom Characteristics	
Soil Amendments	
Diseases & Control	
Pests & Control	

Notes:

Ornamental Plant Files

Common Name	
Latin Name	
Cultivar	
Type	
Bloom Time	
Fragrance	
Source	
Light Preference	
Soil Type	
Water Requirement	
Bloom Characteristics	
Soil Amendments	
Diseases & Control	
Pests & Control	

Notes:

Ornamental Plant Files

Common Name	
Latin Name	
Cultivar	
Type	
Bloom Time	
Fragrance	
Source	
Light Preference	
Soil Type	
Water Requirement	
Bloom Characteristics	
Soil Amendments	
Diseases & Control	
Pests & Control	

Notes:

Ornamental Plant Files

Common Name	
Latin Name	
Cultivar	
Type	
Bloom Time	
Fragrance	
Source	
Light Preference	
Soil Type	
Water Requirement	
Bloom Characteristics	
Soil Amendments	
Diseases & Control	
Pests & Control	

Notes:

Ornamental Plant Files

Common Name	
Latin Name	
Cultivar	
Type	
Bloom Time	
Fragrance	
Source	
Light Preference	
Soil Type	
Water Requirement	
Bloom Characteristics	
Soil Amendments	
Diseases & Control	
Pests & Control	

Notes:

Ornamental Plant Files

Common Name	
Latin Name	
Cultivar	
Type	
Bloom Time	
Fragrance	
Source	
Light Preference	
Soil Type	
Water Requirement	
Bloom Characteristics	
Soil Amendments	
Diseases & Control	
Pests & Control	

Notes:

Ornamental Plant Files

Common Name	
Latin Name	
Cultivar	
Type	
Bloom Time	
Fragrance	
Source	
Light Preference	
Soil Type	
Water Requirement	
Bloom Characteristics	
Soil Amendments	
Diseases & Control	
Pests & Control	

Notes:

Ornamental Plant Files

Common Name	
Latin Name	
Cultivar	
Type	
Bloom Time	
Fragrance	
Source	
Light Preference	
Soil Type	
Water Requirement	
Bloom Characteristics	
Soil Amendments	
Diseases & Control	
Pests & Control	

Notes:

Ornamental Plant Files

Common Name	
Latin Name	
Cultivar	
Type	
Bloom Time	
Fragrance	
Source	
Light Preference	
Soil Type	
Water Requirement	
Bloom Characteristics	
Soil Amendments	
Diseases & Control	
Pests & Control	

Notes:

Ornamental Plant Files

Common Name	
Latin Name	
Cultivar	
Type	
Bloom Time	
Fragrance	
Source	
Light Preference	
Soil Type	
Water Requirement	
Bloom Characteristics	
Soil Amendments	
Diseases & Control	
Pests & Control	

Notes:

Ornamental Plant Files

Common Name	
Latin Name	
Cultivar	
Type	
Bloom Time	
Fragrance	
Source	
Light Preference	
Soil Type	
Water Requirement	
Bloom Characteristics	
Soil Amendments	
Diseases & Control	
Pests & Control	

Notes:

Ornamental Plant Files

Common Name	
Latin Name	
Cultivar	
Type	
Bloom Time	
Fragrance	
Source	
Light Preference	
Soil Type	
Water Requirement	
Bloom Characteristics	
Soil Amendments	
Diseases & Control	
Pests & Control	

Notes:

Notes

Section 4:

Ornamental Plant Files

Common Name	
Latin Name	
Cultivar	
Type	
Bloom Time	
Fragrance	
Source	
Light Preference	
Soil Type	
Water Requirement	
Bloom Characteristics	
Soil Amendments	
Diseases & Control	
Pests & Control	

Notes:

Ornamental Plant Files

Common Name	
Latin Name	
Cultivar	
Type	
Bloom Time	
Fragrance	
Source	
Light Preference	
Soil Type	
Water Requirement	
Bloom Characteristics	
Soil Amendments	
Diseases & Control	
Pests & Control	

Notes:

Ornamental Plant Files

Common Name	
Latin Name	
Cultivar	
Type	
Bloom Time	
Fragrance	
Source	
Light Preference	
Soil Type	
Water Requirement	
Bloom Characteristics	
Soil Amendments	
Diseases & Control	
Pests & Control	

Notes:

Ornamental Plant Files

Common Name	
Latin Name	
Cultivar	
Type	
Bloom Time	
Fragrance	
Source	
Light Preference	
Soil Type	
Water Requirement	
Bloom Characteristics	
Soil Amendments	
Diseases & Control	
Pests & Control	

Notes:

Ornamental Plant Files

Common Name	
Latin Name	
Cultivar	
Type	
Bloom Time	
Fragrance	
Source	
Light Preference	
Soil Type	
Water Requirement	
Bloom Characteristics	
Soil Amendments	
Diseases & Control	
Pests & Control	

Notes:

Ornamental Plant Files

Common Name	
Latin Name	
Cultivar	
Type	
Bloom Time	
Fragrance	
Source	
Light Preference	
Soil Type	
Water Requirement	
Bloom Characteristics	
Soil Amendments	
Diseases & Control	
Pests & Control	

Notes:

Ornamental Plant Files

Common Name	
Latin Name	
Cultivar	
Type	
Bloom Time	
Fragrance	
Source	
Light Preference	
Soil Type	
Water Requirement	
Bloom Characteristics	
Soil Amendments	
Diseases & Control	
Pests & Control	

Notes:

Ornamental Plant Files

Common Name	
Latin Name	
Cultivar	
Type	
Bloom Time	
Fragrance	
Source	
Light Preference	
Soil Type	
Water Requirement	
Bloom Characteristics	
Soil Amendments	
Diseases & Control	
Pests & Control	

Notes:

Ornamental Plant Files

Common Name	
Latin Name	
Cultivar	
Type	
Bloom Time	
Fragrance	
Source	
Light Preference	
Soil Type	
Water Requirement	
Bloom Characteristics	
Soil Amendments	
Diseases & Control	
Pests & Control	

Notes:

Ornamental Plant Files

Common Name	
Latin Name	
Cultivar	
Type	
Bloom Time	
Fragrance	
Source	
Light Preference	
Soil Type	
Water Requirement	
Bloom Characteristics	
Soil Amendments	
Diseases & Control	
Pests & Control	

Notes:

Ornamental Plant Files

Common Name	
Latin Name	
Cultivar	
Type	
Bloom Time	
Fragrance	
Source	
Light Preference	
Soil Type	
Water Requirement	
Bloom Characteristics	
Soil Amendments	
Diseases & Control	
Pests & Control	

Notes:

Ornamental Plant Files

Common Name	
Latin Name	
Cultivar	
Type	
Bloom Time	
Fragrance	
Source	
Light Preference	
Soil Type	
Water Requirement	
Bloom Characteristics	
Soil Amendments	
Diseases & Control	
Pests & Control	

Notes:

Ornamental Plant Files

Common Name	
Latin Name	
Cultivar	
Type	
Bloom Time	
Fragrance	
Source	
Light Preference	
Soil Type	
Water Requirement	
Bloom Characteristics	
Soil Amendments	
Diseases & Control	
Pests & Control	

Notes:

Ornamental Plant Files

Common Name	
Latin Name	
Cultivar	
Type	
Bloom Time	
Fragrance	
Source	
Light Preference	
Soil Type	
Water Requirement	
Bloom Characteristics	
Soil Amendments	
Diseases & Control	
Pests & Control	

Notes:

Notes

Section 5:

Ornamental Plant Files

Common Name	
Latin Name	
Cultivar	
Type	
Bloom Time	
Fragrance	
Source	
Light Preference	
Soil Type	
Water Requirement	
Bloom Characteristics	
Soil Amendments	
Diseases & Control	
Pests & Control	

Notes:

Ornamental Plant Files

Common Name	
Latin Name	
Cultivar	
Type	
Bloom Time	
Fragrance	
Source	
Light Preference	
Soil Type	
Water Requirement	
Bloom Characteristics	
Soil Amendments	
Diseases & Control	
Pests & Control	

Notes:

Ornamental Plant Files

Common Name	
Latin Name	
Cultivar	
Type	
Bloom Time	
Fragrance	
Source	
Light Preference	
Soil Type	
Water Requirement	
Bloom Characteristics	
Soil Amendments	
Diseases & Control	
Pests & Control	

Notes:

Ornamental Plant Files

Common Name	
Latin Name	
Cultivar	
Type	
Bloom Time	
Fragrance	
Source	
Light Preference	
Soil Type	
Water Requirement	
Bloom Characteristics	
Soil Amendments	
Diseases & Control	
Pests & Control	

Notes:

Ornamental Plant Files

Common Name	
Latin Name	
Cultivar	
Type	
Bloom Time	
Fragrance	
Source	
Light Preference	
Soil Type	
Water Requirement	
Bloom Characteristics	
Soil Amendments	
Diseases & Control	
Pests & Control	

Notes:

Ornamental Plant Files

Common Name	
Latin Name	
Cultivar	
Type	
Bloom Time	
Fragrance	
Source	
Light Preference	
Soil Type	
Water Requirement	
Bloom Characteristics	
Soil Amendments	
Diseases & Control	
Pests & Control	

Notes:

Ornamental Plant Files

Common Name	
Latin Name	
Cultivar	
Type	
Bloom Time	
Fragrance	
Source	
Light Preference	
Soil Type	
Water Requirement	
Bloom Characteristics	
Soil Amendments	
Diseases & Control	
Pests & Control	

Notes:

Ornamental Plant Files

Common Name	
Latin Name	
Cultivar	
Type	
Bloom Time	
Fragrance	
Source	
Light Preference	
Soil Type	
Water Requirement	
Bloom Characteristics	
Soil Amendments	
Diseases & Control	
Pests & Control	

Notes:

Ornamental Plant Files

Common Name	
Latin Name	
Cultivar	
Type	
Bloom Time	
Fragrance	
Source	
Light Preference	
Soil Type	
Water Requirement	
Bloom Characteristics	
Soil Amendments	
Diseases & Control	
Pests & Control	

Notes:

Ornamental Plant Files

Common Name	
Latin Name	
Cultivar	
Type	
Bloom Time	
Fragrance	
Source	
Light Preference	
Soil Type	
Water Requirement	
Bloom Characteristics	
Soil Amendments	
Diseases & Control	
Pests & Control	

Notes:

Ornamental Plant Files

Common Name	
Latin Name	
Cultivar	
Type	
Bloom Time	
Fragrance	
Source	
Light Preference	
Soil Type	
Water Requirement	
Bloom Characteristics	
Soil Amendments	
Diseases & Control	
Pests & Control	

Notes:

Ornamental Plant Files

Common Name	
Latin Name	
Cultivar	
Type	
Bloom Time	
Fragrance	
Source	
Light Preference	
Soil Type	
Water Requirement	
Bloom Characteristics	
Soil Amendments	
Diseases & Control	
Pests & Control	

Notes:

Ornamental Plant Files

Common Name	
Latin Name	
Cultivar	
Type	
Bloom Time	
Fragrance	
Source	
Light Preference	
Soil Type	
Water Requirement	
Bloom Characteristics	
Soil Amendments	
Diseases & Control	
Pests & Control	

Notes:

Ornamental Plant Files

Common Name	
Latin Name	
Cultivar	
Type	
Bloom Time	
Fragrance	
Source	
Light Preference	
Soil Type	
Water Requirement	
Bloom Characteristics	
Soil Amendments	
Diseases & Control	
Pests & Control	

Notes:

Notes

Section 6:

Ornamental Plant Files

Common Name	
Latin Name	
Cultivar	
Type	
Bloom Time	
Fragrance	
Source	
Light Preference	
Soil Type	
Water Requirement	
Bloom Characteristics	
Soil Amendments	
Diseases & Control	
Pests & Control	

Notes:

Ornamental Plant Files

Common Name	
Latin Name	
Cultivar	
Type	
Bloom Time	
Fragrance	
Source	
Light Preference	
Soil Type	
Water Requirement	
Bloom Characteristics	
Soil Amendments	
Diseases & Control	
Pests & Control	

Notes:

Ornamental Plant Files

Common Name	
Latin Name	
Cultivar	
Type	
Bloom Time	
Fragrance	
Source	
Light Preference	
Soil Type	
Water Requirement	
Bloom Characteristics	
Soil Amendments	
Diseases & Control	
Pests & Control	

Notes:

Ornamental Plant Files

Common Name	
Latin Name	
Cultivar	
Type	
Bloom Time	
Fragrance	
Source	
Light Preference	
Soil Type	
Water Requirement	
Bloom Characteristics	
Soil Amendments	
Diseases & Control	
Pests & Control	

Notes:

Ornamental Plant Files

Common Name	
Latin Name	
Cultivar	
Type	
Bloom Time	
Fragrance	
Source	
Light Preference	
Soil Type	
Water Requirement	
Bloom Characteristics	
Soil Amendments	
Diseases & Control	
Pests & Control	

Notes:

Ornamental Plant Files

Common Name	
Latin Name	
Cultivar	
Type	
Bloom Time	
Fragrance	
Source	
Light Preference	
Soil Type	
Water Requirement	
Bloom Characteristics	
Soil Amendments	
Diseases & Control	
Pests & Control	

Notes:

Ornamental Plant Files

Common Name	
Latin Name	
Cultivar	
Type	
Bloom Time	
Fragrance	
Source	
Light Preference	
Soil Type	
Water Requirement	
Bloom Characteristics	
Soil Amendments	
Diseases & Control	
Pests & Control	

Notes:

Ornamental Plant Files

Common Name	
Latin Name	
Cultivar	
Type	
Bloom Time	
Fragrance	
Source	
Light Preference	
Soil Type	
Water Requirement	
Bloom Characteristics	
Soil Amendments	
Diseases & Control	
Pests & Control	

Notes:

Ornamental Plant Files

Common Name	
Latin Name	
Cultivar	
Type	
Bloom Time	
Fragrance	
Source	
Light Preference	
Soil Type	
Water Requirement	
Bloom Characteristics	
Soil Amendments	
Diseases & Control	
Pests & Control	

Notes:

Ornamental Plant Files

Common Name	
Latin Name	
Cultivar	
Type	
Bloom Time	
Fragrance	
Source	
Light Preference	
Soil Type	
Water Requirement	
Bloom Characteristics	
Soil Amendments	
Diseases & Control	
Pests & Control	

Notes:

Ornamental Plant Files

Common Name	
Latin Name	
Cultivar	
Type	
Bloom Time	
Fragrance	
Source	
Light Preference	
Soil Type	
Water Requirement	
Bloom Characteristics	
Soil Amendments	
Diseases & Control	
Pests & Control	

Notes:

Ornamental Plant Files

Common Name	
Latin Name	
Cultivar	
Type	
Bloom Time	
Fragrance	
Source	
Light Preference	
Soil Type	
Water Requirement	
Bloom Characteristics	
Soil Amendments	
Diseases & Control	
Pests & Control	

Notes:

Ornamental Plant Files

Common Name	
Latin Name	
Cultivar	
Type	
Bloom Time	
Fragrance	
Source	
Light Preference	
Soil Type	
Water Requirement	
Bloom Characteristics	
Soil Amendments	
Diseases & Control	
Pests & Control	

Notes:

Ornamental Plant Files

Common Name	
Latin Name	
Cultivar	
Type	
Bloom Time	
Fragrance	
Source	
Light Preference	
Soil Type	
Water Requirement	
Bloom Characteristics	
Soil Amendments	
Diseases & Control	
Pests & Control	

Notes:

Notes

Section 7:

Ornamental Plant Files

Common Name	
Latin Name	
Cultivar	
Type	
Bloom Time	
Fragrance	
Source	
Light Preference	
Soil Type	
Water Requirement	
Bloom Characteristics	
Soil Amendments	
Diseases & Control	
Pests & Control	

Notes:

Ornamental Plant Files

Common Name	
Latin Name	
Cultivar	
Type	
Bloom Time	
Fragrance	
Source	
Light Preference	
Soil Type	
Water Requirement	
Bloom Characteristics	
Soil Amendments	
Diseases & Control	
Pests & Control	

Notes:

Ornamental Plant Files

Common Name	
Latin Name	
Cultivar	
Type	
Bloom Time	
Fragrance	
Source	
Light Preference	
Soil Type	
Water Requirement	
Bloom Characteristics	
Soil Amendments	
Diseases & Control	
Pests & Control	

Notes:

Ornamental Plant Files

Common Name	
Latin Name	
Cultivar	
Type	
Bloom Time	
Fragrance	
Source	
Light Preference	
Soil Type	
Water Requirement	
Bloom Characteristics	
Soil Amendments	
Diseases & Control	
Pests & Control	

Notes:

Ornamental Plant Files

Common Name	
Latin Name	
Cultivar	
Type	
Bloom Time	
Fragrance	
Source	
Light Preference	
Soil Type	
Water Requirement	
Bloom Characteristics	
Soil Amendments	
Diseases & Control	
Pests & Control	

Notes:

Ornamental Plant Files

Common Name	
Latin Name	
Cultivar	
Type	
Bloom Time	
Fragrance	
Source	
Light Preference	
Soil Type	
Water Requirement	
Bloom Characteristics	
Soil Amendments	
Diseases & Control	
Pests & Control	

Notes:

Ornamental Plant Files

Common Name	
Latin Name	
Cultivar	
Type	
Bloom Time	
Fragrance	
Source	
Light Preference	
Soil Type	
Water Requirement	
Bloom Characteristics	
Soil Amendments	
Diseases & Control	
Pests & Control	

Notes:

Ornamental Plant Files

Common Name	
Latin Name	
Cultivar	
Type	
Bloom Time	
Fragrance	
Source	
Light Preference	
Soil Type	
Water Requirement	
Bloom Characteristics	
Soil Amendments	
Diseases & Control	
Pests & Control	

Notes:

Notes

Section 8:

Ornamental Plant Files

Common Name	
Latin Name	
Cultivar	
Type	
Bloom Time	
Fragrance	
Source	
Light Preference	
Soil Type	
Water Requirement	
Bloom Characteristics	
Soil Amendments	
Diseases & Control	
Pests & Control	

Notes:

Ornamental Plant Files

Common Name	
Latin Name	
Cultivar	
Type	
Bloom Time	
Fragrance	
Source	
Light Preference	
Soil Type	
Water Requirement	
Bloom Characteristics	
Soil Amendments	
Diseases & Control	
Pests & Control	

Notes:

Ornamental Plant Files

Common Name	
Latin Name	
Cultivar	
Type	
Bloom Time	
Fragrance	
Source	
Light Preference	
Soil Type	
Water Requirement	
Bloom Characteristics	
Soil Amendments	
Diseases & Control	
Pests & Control	

Notes:

Ornamental Plant Files

Common Name	
Latin Name	
Cultivar	
Type	
Bloom Time	
Fragrance	
Source	
Light Preference	
Soil Type	
Water Requirement	
Bloom Characteristics	
Soil Amendments	
Diseases & Control	
Pests & Control	

Notes:

Ornamental Plant Files

Common Name	
Latin Name	
Cultivar	
Type	
Bloom Time	
Fragrance	
Source	
Light Preference	
Soil Type	
Water Requirement	
Bloom Characteristics	
Soil Amendments	
Diseases & Control	
Pests & Control	

Notes:

Ornamental Plant Files

Common Name	
Latin Name	
Cultivar	
Type	
Bloom Time	
Fragrance	
Source	
Light Preference	
Soil Type	
Water Requirement	
Bloom Characteristics	
Soil Amendments	
Diseases & Control	
Pests & Control	

Notes:

Ornamental Plant Files

Common Name	
Latin Name	
Cultivar	
Type	
Bloom Time	
Fragrance	
Source	
Light Preference	
Soil Type	
Water Requirement	
Bloom Characteristics	
Soil Amendments	
Diseases & Control	
Pests & Control	

Notes:

Ornamental Plant Files

Common Name	
Latin Name	
Cultivar	
Type	
Bloom Time	
Fragrance	
Source	
Light Preference	
Soil Type	
Water Requirement	
Bloom Characteristics	
Soil Amendments	
Diseases & Control	
Pests & Control	

Notes:

Ornamental Plant Files

Common Name	
Latin Name	
Cultivar	
Type	
Bloom Time	
Fragrance	
Source	
Light Preference	
Soil Type	
Water Requirement	
Bloom Characteristics	
Soil Amendments	
Diseases & Control	
Pests & Control	

Notes:

Ornamental Plant Files

Common Name	
Latin Name	
Cultivar	
Type	
Bloom Time	
Fragrance	
Source	
Light Preference	
Soil Type	
Water Requirement	
Bloom Characteristics	
Soil Amendments	
Diseases & Control	
Pests & Control	

Notes:

Ornamental Plant Files

Common Name	
Latin Name	
Cultivar	
Type	
Bloom Time	
Fragrance	
Source	
Light Preference	
Soil Type	
Water Requirement	
Bloom Characteristics	
Soil Amendments	
Diseases & Control	
Pests & Control	

Notes:

Ornamental Plant Files

Common Name	
Latin Name	
Cultivar	
Type	
Bloom Time	
Fragrance	
Source	
Light Preference	
Soil Type	
Water Requirement	
Bloom Characteristics	
Soil Amendments	
Diseases & Control	
Pests & Control	

Notes:

Ornamental Plant Files

Common Name	
Latin Name	
Cultivar	
Type	
Bloom Time	
Fragrance	
Source	
Light Preference	
Soil Type	
Water Requirement	
Bloom Characteristics	
Soil Amendments	
Diseases & Control	
Pests & Control	

Notes:

Ornamental Plant Files

Common Name	
Latin Name	
Cultivar	
Type	
Bloom Time	
Fragrance	
Source	
Light Preference	
Soil Type	
Water Requirement	
Bloom Characteristics	
Soil Amendments	
Diseases & Control	
Pests & Control	

Notes:

Notes

Section 9:

Ornamental Plant Files

Common Name	
Latin Name	
Cultivar	
Type	
Bloom Time	
Fragrance	
Source	
Light Preference	
Soil Type	
Water Requirement	
Bloom Characteristics	
Soil Amendments	
Diseases & Control	
Pests & Control	

Notes:

Ornamental Plant Files

Common Name	
Latin Name	
Cultivar	
Type	
Bloom Time	
Fragrance	
Source	
Light Preference	
Soil Type	
Water Requirement	
Bloom Characteristics	
Soil Amendments	
Diseases & Control	
Pests & Control	

Notes:

Ornamental Plant Files

Common Name	
Latin Name	
Cultivar	
Type	
Bloom Time	
Fragrance	
Source	
Light Preference	
Soil Type	
Water Requirement	
Bloom Characteristics	
Soil Amendments	
Diseases & Control	
Pests & Control	

Notes:

Ornamental Plant Files

Common Name	
Latin Name	
Cultivar	
Type	
Bloom Time	
Fragrance	
Source	
Light Preference	
Soil Type	
Water Requirement	
Bloom Characteristics	
Soil Amendments	
Diseases & Control	
Pests & Control	

Notes:

Ornamental Plant Files

Common Name	
Latin Name	
Cultivar	
Type	
Bloom Time	
Fragrance	
Source	
Light Preference	
Soil Type	
Water Requirement	
Bloom Characteristics	
Soil Amendments	
Diseases & Control	
Pests & Control	

Notes:

Ornamental Plant Files

Common Name	
Latin Name	
Cultivar	
Type	
Bloom Time	
Fragrance	
Source	
Light Preference	
Soil Type	
Water Requirement	
Bloom Characteristics	
Soil Amendments	
Diseases & Control	
Pests & Control	

Notes:

Ornamental Plant Files

Common Name	
Latin Name	
Cultivar	
Type	
Bloom Time	
Fragrance	
Source	
Light Preference	
Soil Type	
Water Requirement	
Bloom Characteristics	
Soil Amendments	
Diseases & Control	
Pests & Control	

Notes:

Ornamental Plant Files

Common Name	
Latin Name	
Cultivar	
Type	
Bloom Time	
Fragrance	
Source	
Light Preference	
Soil Type	
Water Requirement	
Bloom Characteristics	
Soil Amendments	
Diseases & Control	
Pests & Control	

Notes:

Ornamental Plant Files

Common Name	
Latin Name	
Cultivar	
Type	
Bloom Time	
Fragrance	
Source	
Light Preference	
Soil Type	
Water Requirement	
Bloom Characteristics	
Soil Amendments	
Diseases & Control	
Pests & Control	

Notes:

Ornamental Plant Files

Common Name	
Latin Name	
Cultivar	
Type	
Bloom Time	
Fragrance	
Source	
Light Preference	
Soil Type	
Water Requirement	
Bloom Characteristics	
Soil Amendments	
Diseases & Control	
Pests & Control	

Notes:

Notes

Section 10:

Ornamental Plant Files

Common Name	
Latin Name	
Cultivar	
Type	
Bloom Time	
Fragrance	
Source	
Light Preference	
Soil Type	
Water Requirement	
Bloom Characteristics	
Soil Amendments	
Diseases & Control	
Pests & Control	

Notes:

Ornamental Plant Files

Common Name	
Latin Name	
Cultivar	
Type	
Bloom Time	
Fragrance	
Source	
Light Preference	
Soil Type	
Water Requirement	
Bloom Characteristics	
Soil Amendments	
Diseases & Control	
Pests & Control	

Notes:

Ornamental Plant Files

Common Name	
Latin Name	
Cultivar	
Type	
Bloom Time	
Fragrance	
Source	
Light Preference	
Soil Type	
Water Requirement	
Bloom Characteristics	
Soil Amendments	
Diseases & Control	
Pests & Control	

Notes:

Ornamental Plant Files

Common Name	
Latin Name	
Cultivar	
Type	
Bloom Time	
Fragrance	
Source	
Light Preference	
Soil Type	
Water Requirement	
Bloom Characteristics	
Soil Amendments	
Diseases & Control	
Pests & Control	

Notes:

Ornamental Plant Files

Common Name	
Latin Name	
Cultivar	
Type	
Bloom Time	
Fragrance	
Source	
Light Preference	
Soil Type	
Water Requirement	
Bloom Characteristics	
Soil Amendments	
Diseases & Control	
Pests & Control	

Notes:

Ornamental Plant Files

Common Name	
Latin Name	
Cultivar	
Type	
Bloom Time	
Fragrance	
Source	
Light Preference	
Soil Type	
Water Requirement	
Bloom Characteristics	
Soil Amendments	
Diseases & Control	
Pests & Control	

Notes:

Ornamental Plant Files

Common Name	
Latin Name	
Cultivar	
Type	
Bloom Time	
Fragrance	
Source	
Light Preference	
Soil Type	
Water Requirement	
Bloom Characteristics	
Soil Amendments	
Diseases & Control	
Pests & Control	

Notes:

Ornamental Plant Files

Common Name	
Latin Name	
Cultivar	
Type	
Bloom Time	
Fragrance	
Source	
Light Preference	
Soil Type	
Water Requirement	
Bloom Characteristics	
Soil Amendments	
Diseases & Control	
Pests & Control	

Notes:

Notes

Section 11:

Ornamental Plant Files

Common Name	
Latin Name	
Cultivar	
Type	
Bloom Time	
Fragrance	
Source	
Light Preference	
Soil Type	
Water Requirement	
Bloom Characteristics	
Soil Amendments	
Diseases & Control	
Pests & Control	

Notes:

Ornamental Plant Files

Common Name	
Latin Name	
Cultivar	
Type	
Bloom Time	
Fragrance	
Source	
Light Preference	
Soil Type	
Water Requirement	
Bloom Characteristics	
Soil Amendments	
Diseases & Control	
Pests & Control	

Notes:

Ornamental Plant Files

Common Name	
Latin Name	
Cultivar	
Type	
Bloom Time	
Fragrance	
Source	
Light Preference	
Soil Type	
Water Requirement	
Bloom Characteristics	
Soil Amendments	
Diseases & Control	
Pests & Control	

Notes:

Ornamental Plant Files

Common Name	
Latin Name	
Cultivar	
Type	
Bloom Time	
Fragrance	
Source	
Light Preference	
Soil Type	
Water Requirement	
Bloom Characteristics	
Soil Amendments	
Diseases & Control	
Pests & Control	

Notes:

Ornamental Plant Files

Common Name	
Latin Name	
Cultivar	
Type	
Bloom Time	
Fragrance	
Source	
Light Preference	
Soil Type	
Water Requirement	
Bloom Characteristics	
Soil Amendments	
Diseases & Control	
Pests & Control	

Notes:

Ornamental Plant Files

Common Name	
Latin Name	
Cultivar	
Type	
Bloom Time	
Fragrance	
Source	
Light Preference	
Soil Type	
Water Requirement	
Bloom Characteristics	
Soil Amendments	
Diseases & Control	
Pests & Control	

Notes:

Ornamental Plant Files

Common Name	
Latin Name	
Cultivar	
Type	
Bloom Time	
Fragrance	
Source	
Light Preference	
Soil Type	
Water Requirement	
Bloom Characteristics	
Soil Amendments	
Diseases & Control	
Pests & Control	

Notes:

Ornamental Plant Files

Common Name	
Latin Name	
Cultivar	
Type	
Bloom Time	
Fragrance	
Source	
Light Preference	
Soil Type	
Water Requirement	
Bloom Characteristics	
Soil Amendments	
Diseases & Control	
Pests & Control	

Notes:

Notes

Section 12:

Ornamental Plant Files

Common Name	
Latin Name	
Cultivar	
Type	
Bloom Time	
Fragrance	
Source	
Light Preference	
Soil Type	
Water Requirement	
Bloom Characteristics	
Soil Amendments	
Diseases & Control	
Pests & Control	

Notes:

Ornamental Plant Files

Common Name	
Latin Name	
Cultivar	
Type	
Bloom Time	
Fragrance	
Source	
Light Preference	
Soil Type	
Water Requirement	
Bloom Characteristics	
Soil Amendments	
Diseases & Control	
Pests & Control	

Notes:

Ornamental Plant Files

Common Name	
Latin Name	
Cultivar	
Type	
Bloom Time	
Fragrance	
Source	
Light Preference	
Soil Type	
Water Requirement	
Bloom Characteristics	
Soil Amendments	
Diseases & Control	
Pests & Control	

Notes:

Ornamental Plant Files

Common Name	
Latin Name	
Cultivar	
Type	
Bloom Time	
Fragrance	
Source	
Light Preference	
Soil Type	
Water Requirement	
Bloom Characteristics	
Soil Amendments	
Diseases & Control	
Pests & Control	

Notes:

Ornamental Plant Files

Common Name	
Latin Name	
Cultivar	
Type	
Bloom Time	
Fragrance	
Source	
Light Preference	
Soil Type	
Water Requirement	
Bloom Characteristics	
Soil Amendments	
Diseases & Control	
Pests & Control	

Notes:

Ornamental Plant Files

Common Name	
Latin Name	
Cultivar	
Type	
Bloom Time	
Fragrance	
Source	
Light Preference	
Soil Type	
Water Requirement	
Bloom Characteristics	
Soil Amendments	
Diseases & Control	
Pests & Control	

Notes:

Ornamental Plant Files

Common Name	
Latin Name	
Cultivar	
Type	
Bloom Time	
Fragrance	
Source	
Light Preference	
Soil Type	
Water Requirement	
Bloom Characteristics	
Soil Amendments	
Diseases & Control	
Pests & Control	

Notes:

Ornamental Plant Files

Common Name	
Latin Name	
Cultivar	
Type	
Bloom Time	
Fragrance	
Source	
Light Preference	
Soil Type	
Water Requirement	
Bloom Characteristics	
Soil Amendments	
Diseases & Control	
Pests & Control	

Notes:

Notes

Section 13:

Ornamental Plant Files

Common Name	
Latin Name	
Cultivar	
Type	
Bloom Time	
Fragrance	
Source	
Light Preference	
Soil Type	
Water Requirement	
Bloom Characteristics	
Soil Amendments	
Diseases & Control	
Pests & Control	

Notes:

Ornamental Plant Files

Common Name	
Latin Name	
Cultivar	
Type	
Bloom Time	
Fragrance	
Source	
Light Preference	
Soil Type	
Water Requirement	
Bloom Characteristics	
Soil Amendments	
Diseases & Control	
Pests & Control	

Notes:

Ornamental Plant Files

Common Name	
Latin Name	
Cultivar	
Type	
Bloom Time	
Fragrance	
Source	
Light Preference	
Soil Type	
Water Requirement	
Bloom Characteristics	
Soil Amendments	
Diseases & Control	
Pests & Control	

Notes:

Ornamental Plant Files

Common Name	
Latin Name	
Cultivar	
Type	
Bloom Time	
Fragrance	
Source	
Light Preference	
Soil Type	
Water Requirement	
Bloom Characteristics	
Soil Amendments	
Diseases & Control	
Pests & Control	

Notes:

Ornamental Plant Files

Common Name	
Latin Name	
Cultivar	
Type	
Bloom Time	
Fragrance	
Source	
Light Preference	
Soil Type	
Water Requirement	
Bloom Characteristics	
Soil Amendments	
Diseases & Control	
Pests & Control	

Notes:

Ornamental Plant Files

Common Name	
Latin Name	
Cultivar	
Type	
Bloom Time	
Fragrance	
Source	
Light Preference	
Soil Type	
Water Requirement	
Bloom Characteristics	
Soil Amendments	
Diseases & Control	
Pests & Control	

Notes:

Notes

Section 14:

Ornamental Plant Files

Common Name	
Latin Name	
Cultivar	
Type	
Bloom Time	
Fragrance	
Source	
Light Preference	
Soil Type	
Water Requirement	
Bloom Characteristics	
Soil Amendments	
Diseases & Control	
Pests & Control	

Notes:

Ornamental Plant Files

Common Name	
Latin Name	
Cultivar	
Type	
Bloom Time	
Fragrance	
Source	
Light Preference	
Soil Type	
Water Requirement	
Bloom Characteristics	
Soil Amendments	
Diseases & Control	
Pests & Control	

Notes:

Ornamental Plant Files

Common Name	
Latin Name	
Cultivar	
Type	
Bloom Time	
Fragrance	
Source	
Light Preference	
Soil Type	
Water Requirement	
Bloom Characteristics	
Soil Amendments	
Diseases & Control	
Pests & Control	

Notes:

Ornamental Plant Files

Common Name	
Latin Name	
Cultivar	
Type	
Bloom Time	
Fragrance	
Source	
Light Preference	
Soil Type	
Water Requirement	
Bloom Characteristics	
Soil Amendments	
Diseases & Control	
Pests & Control	

Notes:

Ornamental Plant Files

Common Name	
Latin Name	
Cultivar	
Type	
Bloom Time	
Fragrance	
Source	
Light Preference	
Soil Type	
Water Requirement	
Bloom Characteristics	
Soil Amendments	
Diseases & Control	
Pests & Control	

Notes:

Ornamental Plant Files

Common Name	
Latin Name	
Cultivar	
Type	
Bloom Time	
Fragrance	
Source	
Light Preference	
Soil Type	
Water Requirement	
Bloom Characteristics	
Soil Amendments	
Diseases & Control	
Pests & Control	

Notes:

Notes

Section 15:

Ornamental Plant Files

Common Name	
Latin Name	
Cultivar	
Type	
Bloom Time	
Fragrance	
Source	
Light Preference	
Soil Type	
Water Requirement	
Bloom Characteristics	
Soil Amendments	
Diseases & Control	
Pests & Control	

Notes:

Ornamental Plant Files

Common Name	
Latin Name	
Cultivar	
Type	
Bloom Time	
Fragrance	
Source	
Light Preference	
Soil Type	
Water Requirement	
Bloom Characteristics	
Soil Amendments	
Diseases & Control	
Pests & Control	

Notes:

Ornamental Plant Files

Common Name	
Latin Name	
Cultivar	
Type	
Bloom Time	
Fragrance	
Source	
Light Preference	
Soil Type	
Water Requirement	
Bloom Characteristics	
Soil Amendments	
Diseases & Control	
Pests & Control	

Notes:

Ornamental Plant Files

Common Name	
Latin Name	
Cultivar	
Type	
Bloom Time	
Fragrance	
Source	
Light Preference	
Soil Type	
Water Requirement	
Bloom Characteristics	
Soil Amendments	
Diseases & Control	
Pests & Control	

Notes:

Ornamental Plant Files

Common Name	
Latin Name	
Cultivar	
Type	
Bloom Time	
Fragrance	
Source	
Light Preference	
Soil Type	
Water Requirement	
Bloom Characteristics	
Soil Amendments	
Diseases & Control	
Pests & Control	

Notes:

Ornamental Plant Files

Common Name	
Latin Name	
Cultivar	
Type	
Bloom Time	
Fragrance	
Source	
Light Preference	
Soil Type	
Water Requirement	
Bloom Characteristics	
Soil Amendments	
Diseases & Control	
Pests & Control	

Notes:

Notes

Notes

Melanie Holsti is an avid tea lover with a deep appreciation for the simple joys of life. Whether she's sipping a cup of Scottish Breakfast tea, capturing the perfect shot with her camera, or getting her hands dirty in the garden, she embraces each moment with enthusiasm.

As an author and small business owner, she pours her focus and creativity into everything she does.

Her background is as varied as her interests—she's worked in hospitality, retail, livestock farming, writing, and, of course, gardening. Over the years, she has experienced the triumphs and challenges that come with growing food and homesteading in diverse climates, from the harsh prairie of South Dakota to the woods and mountains of Missouri and Tennessee.

In 2018, Melanie and her family took on an exciting project—restoring a charming 1917 Craftsman bungalow. With a lot of hard work and a shared vision, they transformed it into the Ironstone Cottage Tea Room, where guests could gather and enjoy a hot cup of tea or lunch with friends in a cozy, welcoming atmosphere.

These days, Melanie, her husband, and their daughters reside on a small farm in Tennessee, living in a beautiful Federal-style home they built together. Their homestead is shared with an assortment of animals, including a peculiar flock of barn cats, free-ranging chickens, and noisy ducks.

When she's not tending to the garden or caring for her animals, Melanie finds joy in writing books, sampling new and exotic teas, and cheering on her favorite professional hockey team. No matter the season, she's always pursuing her interests and creating a life filled with warmth and, of course, plenty of tea.

www.ironstone tennessee.com

IronstoneTennessee

IronstoneTennessee

amz

amazon.com/author/ironstonetennessee

ironstonetennessee

secret resources web page
Password: IronstoneBooks

Your feedback and reviews are priceless. If you have suggestions for improving my gardening workbooks, I'd love to hear them. If you have a moment, please consider leaving a review on Amazon.

Other Titles by Melanie Holsti

The Ironstone Garden Planner (color)

The Ironstone Garden Journal(color)

The Vegetable Files

The Houseplant Files

The Herb Files

The Orchard Files

The Fruit, Flower, Herb & Vegetable Garden Planner

The Fruit, Flower, Herb & Vegetable Garden Journal

Journaling Your Weather

The Ironstone Forager's Notebook

Journaling Your Health (various designs)

Ironstone Prayer Journals (various designs)

Church & Assembly Prayer Team Journals (various designs)

Tea & Tisane Blending Workbook: A Tea Mixologist's Journal

(with Jennifer C. Petersen)

Made in the USA
Monee, IL
05 November 2026

67911531R10122